MATH and ME

Math in the Kitchen

by Joanne Mattern

LOOK! BOOKS™

Red Chair Press Egremont, Massachusetts

Look! Books are produced and published by Red Chair Press:
Red Chair Press LLC PO Box 333 South Egremont, MA 01258-0333

FREE Educator Guides at www.redchairpress.com/free-resources

Publisher's Cataloging-In-Publication Data
Names: Mattern, Joanne, 1963- author.
Title: Math in the kitchen / by Joanne Mattern.

Description: Egremont, Massachusetts : Red Chair Press, [2022] | Series: LOOK! books : Math and Me | Interest age level: 005-008. | Includes index and suggested resources for further reading. | Summary: "Math isn't just something you learn in school. It's an important part of the world around you. What's better than baking a yummy pizza? Using math skills while you cook! Readers explore concepts such as measuring, adding, dividing, and counting in this cooking adventure"--Provided by publisher.

Identifiers: ISBN 9781643711324 (hardcover) | ISBN 9781643711386 (softcover) | ISBN 9781643711447 (ePDF) | ISBN 9781643711508 (ePub 3 S&L) | ISBN 9781643711560 (ePub 3 TR) | ISBN 9781643711621 (Kindle)

Subjects: LCSH: Mathematics--Juvenile literature. | Cooking--Mathematics--Juvenile literature. | CYAC: Mathematics. | Cooking--Mathematics.

Classification: LCC QA40.5 .M385 2022 (print) | LCC QA40.5 (ebook) | DDC 510--dc23

Library of Congress Control Number: 2021945370

Photo credits: Cover, p. 19: iStock; p. 3–18, 20–24: Shutterstock

Printed in United States of America
0422 1P CGF22

Table of Contents

In the Kitchen

It is fun to make things in the kitchen. Let's bake some tasty cookies! Think about the math you will use while you bake.

How Many?

You need flour, sugar, eggs, baking soda, and chocolate chips to bake your cookies. How many **ingredients** is that? Let's count.

1, 2, 3, 4, 5.

You need 5 ingredients.

You might find directions online or in a book for things you want to cook.

Let's Measure

You need 2 cups of flour. Let's **measure**. This spoon holds ½ cup. How many times should you fill it?

MATH FACT!

½ + ½ = 1

Two half-cups equal one cup. We need two of that amount or 4 half-cups to make 2 cups.

Next you need half a cup of sugar. Is ½ a cup of sugar less than 2 cups of flour? Yes!

MATH FACT!

$\frac{1}{2} < 2$

$\frac{1}{2}$ is less than 2.
2 is greater than $\frac{1}{2}$.

Write this: $2 > \frac{1}{2}$.

Counting Eggs

The **recipe** says you need 2 eggs. There are 6 eggs in the carton. If you use 2 eggs, how many eggs are left?

MATH FACT!

6 – 2 = 4

There are 4 eggs left.

Lots of Chips

Stir in the baking soda. It's time to add the chips! Measure one cup. Then stir them into the **batter**.

Some people use raisins, not chips, for a healthy choice.

Time to Bake

Scoop the batter onto the cookie sheets. Each sheet holds 8 cookies. You can fill 3 sheets. How many cookies will you have?

MATH FACT!

8 + 8 = 16
16 + 8 = 24

8 + 8 + 8 = 24

You have 24 cookies.

The cookies need to bake for 12 minutes. It is 2:00. What time will the cookies be ready?

MATH FACT!

2:00 + 12 minutes = 2:12

Ask an adult to help you when using an oven.

Let's Eat!

The cookies are done! Place them on a rack to cool. Then you can eat them! Let's eat 4 cookies. How many cookies will be left?

MATH FACT!

24 – 4 = 20

There are 20 cookies left.

Kitchen Fun

Your cookies taste so good! It's fun to use math to bake tasty treats. What will you cook next?

Words to Know

batter: a mixture of flour and other ingredients used in baking

ingredients: foods that are put together to make something

measure: to find out how much of something you have

recipe: instructions to make food

Learn More at the Library

Check out these books to learn more.

Steffora, Tracey. *Shapes in the Kitchen.* Heinemann-Raintree, 2011.

Steffora, Tracey. *What is Math?* Heinemann-Raintree, 2012.

Weiss, Ellen. *Math in the Kitchen.* Children's Press, 2007.

Index

About the Author

Joanne Mattern is the author of many books for children. She loves writing about sports, animals, and interesting people. Mattern lives in New York State with her family.